BRIDGING

DISCOVERING THE BEAUTY OF BRIDGES

Chain Bridge, Budapest

BRIDGING

DISCOVERING THE BEAUTY OF BRIDGES

Robert S. Cortright

BRIDGE INK
10200 SW Hoodview Drive
Tigard, Oregon 97224

Library of Congress Catalog Card Number: 98-92665

ISBN: 0-9641963-2-8

BRIDGE INK
10200 SW Hoodview Drive
Tigard, Oregon 97224

Editing, Jeanne Cortright Neff
Imagesetting, Negative Perfection Ltd.
Binding, Lincoln & Allen
Printed in USA by Millcross Litho

TO MY WIFE, KATHY

For her support and encouragement in spite
of having to wait patiently at every bridge.

Pont d'Avignon, France

CONTENTS

Bridging

DISCOVERING THE BEAUTY OF BRIDGES

ridges, we all use them, but we seldom see them. For me, discovering bridges came with the first opportunity to travel abroad. Originally I was an enthusiastic tourist, thrilled with all of the sights encountered in travel. Gradually the focus of attention and the focus of the camera began to be concentrated upon bridges. Ultimately, that concentration escalated to the level of an obsession. From the resulting accumulation of bridge pictures grew a compelling desire to share. That led in 1994 to the publishing of a collection of pictures in a book entitled "*Bridging.*"

Since publication, "*Bridging*" has brought me into contact with a number of readers from near and far. It has been especially gratifying to learn that many who weren't previously conscious of bridges now see and admire the bridges around them. They have discovered the beauty of bridges.

As my travels in search of bridges have continued, my portfolio has grown to over 4,000 pictures of bridges in 20 countries and I want to share the best of them with you. If you have already learned to appreciate bridges, this book should captivate you. If you are one who has previously ignored the bridges you've encountered, this book is intended to heighten your awareness so that you too can discover the beauty of bridges.

BRIDGE TYPES

Before we embark on our journey of bridge discovery, let's examine the types of bridges we will encounter.

BRIDGES OF STONE

The bridges that have survived from ancient times are constructed of stone. The stone arch is not only durable, but a thing of beauty.

Though the ancient Romans were not the first to utilize stone for bridge building, the many wonderful Roman spans still in use today attest to their skill in the design and construction of arched stone bridges.

This outstanding example of Roman construction is at Alcántara, Spain. The bridge was built for Emperor Trajan in 104 A.D. by Caius Julius Lacer. Almost 1,900 years later, the bridge still looks new.

After the decline of the Roman empire, Europe saw no significant new stone bridges until medieval times. Prague's Charles Bridge is a venerable example from the 14th century. Now a pedestrian bridge, it has always played a key role in the history and legends of the city.

The historic bridge at Mostar in Bosnia, built by the Ottoman Turks in 1566, was the cherished centerpiece of this beautiful little town. The tourists, the daredevil and the bridge are gone now, victims of the recent civil war in the former Yugoslavia.

Although this charming bridge in the Croatian village of Proložac is not famous as was the bridge at nearby Mostar, it has special significance to our family. One of the modest stone houses near the bridge was the birthplace of my father-in-law. The bridge appears to be Turkish in style and probably dates from the 16th century.

This bridge in Kortrijk, Belgium also has a family connection. My Cortright (Kortrijk) ancestors assumed the name of Van Kortrijk when they left this Flemish town over 300 years ago and landed in New Amsterdam (New York).

WOODEN BRIDGES

Wood undoubtedly pre-dates stone as a bridge building material.
Even now, some of Rome's bridges are resting on the original
wooden pilings driven into the Tiber riverbed 2,000 years ago.
But wood exposed to the weather does not compare in durability
to stone, iron and steel, or concrete.

Bridges of wood come in a variety of forms,
some ageless in style and some quite modern.

This most unusual bridge is at Queens' College in Cambridge, England. Legend has it that students constructed the wooden "Mathematical" bridge in 1749 on geometric principles and without nails.

A common means of protecting a wooden bridge from the elements is to cover it. My home state of Oregon has many carefully preserved covered bridges such as this one at Chitwood. While many people find this style especially appealing, I think of it as a bridge in a barn.

A trip to Switzerland revised my thinking. I must now admit that covered bridges can be very charming.

The Spreuerbrücke in Luzern has been in service since 1568.

The famous Kapellbrücke is also in Luzern. Built in 1333, it was most recently restored in 1994 after being almost totally destroyed by fire.

Modern wooden bridges often incorporate laminated beams. This graceful arch serves a highway interchange in South Dakota.

The Shady Cove Bridge is a laminated wood tied arch built strong enough to support heavily loaded trucks on this logging road in Oregon.

BRIDGES OF IRON AND STEEL

Iron was used to clamp blocks of stone together in bridges as far back as Roman times. But this material wasn't employed as the main structural element until the 18th century.

In 1779 Abraham Darby's innovative arch over the Severn River in England heralded a new era in bridge construction. This was the first span built entirely of iron. The five 100 foot long cast-iron ribs were assembled in the manner of intricate carpentry. The original structure is still in use but only for pedestrians. It is a major tourist attraction in the town which is now called Ironbridge.

As the use of cast iron in bridges became more common, builders often embellished their creations with elaborate decoration. Thomas Telford built this bridge at Betwys-y-Coed, Wales in 1815. It is worth a climb down the riverbank to enjoy the view of the brightly painted ironwork.

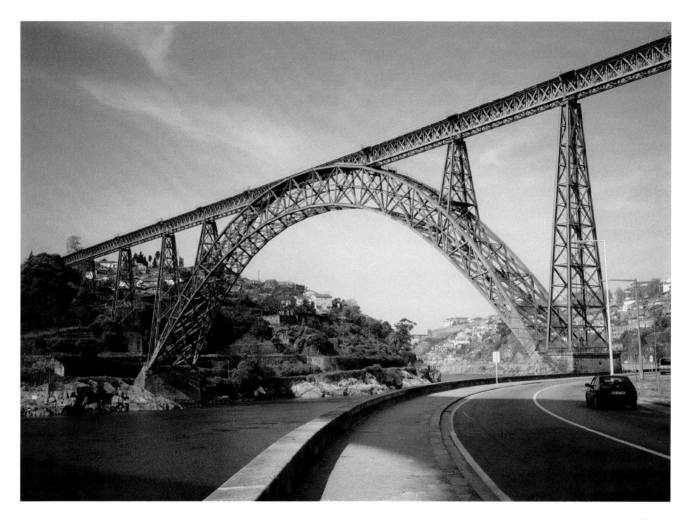

As cast iron was supplanted by wrought iron, more ambitious bridges were attempted. Gustave Eiffel built many bridges before the construction of the famous Eiffel tower. His Ponte Maria Pia in Porto, Portugal was completed in 1877. The graceful arch spanning 525 feet is now idle. For 114 years it carried rail traffic over the Douro River until the completion of the new Sâo Joâo Bridge in 1991 (see page 123.)

England

California

The development of iron and steel bridges provided spans of great strength and economy, but somewhere along the way the aesthetics of bridge design seemed to lose importance. Unattractive examples are only too common.

Portugal

Arizona

When the arch form is chosen it seems that the visual effect is always enhanced. These examples demonstrate variations of the arch form.

The Marechal Carmona Bridge in Portugal is an example of the tied arch. The roadway is suspended from the arches.

The Perrine Bridge spans the Snake River Canyon at Twin Falls, Idaho. This is a deck arch, a form that lends itself to a site where the canyon walls can accept the horizontal thrust of the arch.

The Tyne Bridge in England is a through arch. The deck is partially suspended from the arch and partially supported above the arch. The traffic passes through the arch. (Locally this bridge is called "the coat hanger.")

BRIDGES OF CONCRETE

Concrete containing a volcanic ash called pozzalana was successfully employed as a building material by the Romans. But concrete didn't come into structural use again until the 18th century. Since the late 19th century the development of reinforced and pre-stressed concrete has provided bridge engineers with many new options in bridge design.

Scotland

Oregon

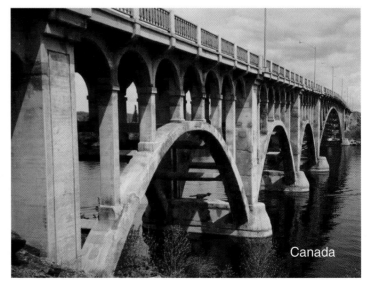

Canada

Alaska

Portugal

Italy

In the Scottish Highlands, the huge Glenfinnan Viaduct is unusual in that it was constructed of mass concrete, that is, without iron or steel reinforcement. Now 100 years old, it still carries trains in a graceful curve 90 feet above the river Finnan.

In the first half of the 20th century, bridge engineers found that the reinforced concrete arch provided the opportunity to produce especially attractive spans. The Rainbow Bridge at Folsom, California reflects the aesthetic appeal of an open spandrel deck arch.

Although the more modern renditions usually lack the decorative embellishments of earlier generations of concrete bridges, the size of their sweeping arches can create a breathtaking scene. The Redman Memorial Bridge at Selah Creek in Washington is one of the largest concrete arches in America. Built in 1971, its twin arches carry the interstate highway 549 feet across the valley.

These huge Croatian spans are among the most impressive in my collection. Until it was surpassed in China in 1996, the bridge above was the longest concrete arch in the world. The span is 1,280 feet. It is the larger of a pair that connect the Island of Krk to the Croatian mainland

The 808 foot span on the right was built in 1967. It carries the coastal highway over a deep inlet near Šibenik.

MOVING BRIDGES

There are two principal ways to avoid obstruction of traffic under a bridge. One is to build a bridge with sufficient vertical clearance to permit traffic to pass unimpeded under the span. When this is impractical, the alternative is to build a bridge that moves. The common forms are the bascule, the swing bridge and the lift bridge.

BASCULE BRIDGES

The bascule employs the same principle as a draw-bridge that was lifted to secure a castle. Most are not likely candidates for inclusion in a book about "the Beauty of Bridges," even with a nice blue paint job, like the example above.

An attractive exception is this charming drawspan which still carries traffic over the canal and into the town gate at Zierikzee, The Netherlands.

SWING BRIDGES   Another way to move a bridge is to swing it out of the way. While this allows unlimited vertical clearance, it tends to create an obstruction to the channel.

Pictured here is the swing bridge at Newcastle-upon-Tyne in England. It was built in 1876. It provides two 110 foot navigation channels when open.

There are three interesting bridges at this location. Dominating the background is the High Level Bridge, built in 1849 by Robert Stephenson. This photograph was taken from the deck of the third, the Tyne Bridge (see page 19).

LIFT BRIDGES   Vertical lift spans provide an efficient but usually ugly means to maintain reasonable traffic flow both on and under bridges.

Tower Bridge in Sacramento, California is a typical configuration. Its 209 foot deck can be lifted high above the channel in a matter of seconds to permit the passage of waterborne traffic. The counterweights are concealed in the towers.

The St. Johns Railroad Bridge in Portland, Oregon was built in 1908 as a swing span. In 1987 it was rebuilt and provided with a 532 foot lift span to improve navigation on the Willamette River.

In this modern attempt to produce an attractive lift bridge, almost all of the lifting mechanism is hidden in the towers of the Green Island Bridge over the Hudson River at Troy, New York.

The transporter bridge is unique among moving bridges. A few of these monsters were built a hundred years ago to enable pedestrian and vehicular traffic to cross busy harbors full of tall ships without obstructing or interrupting seagoing commerce. The "aerial ferry" was faster than a conventional ferry and more reliable in bad weather. But with the explosion of automobile traffic, the carrying capacity was too limited to be practical.

This is the Newport Transporter Bridge in Wales, designed by Ferdinand Arodin, a French engineer who was the leading exponent of the type. The overhead beam is 177 feet above the channel and is actually a suspension bridge with a span of 645 feet. The gondola is suspended by cables from a traveller which is pulled along the beam by cable. It opened for business in 1906, and recently restored, it still carries traffic across the River Usk. It is a very smooth ride.

SUSPENSION BRIDGES

The beauty of the arch form is rivaled by the sweep of the graceful catenary curve of a suspension bridge. There seems to be an endless variety in the design of the towers that support the chains or cables.

While modern bridges are supported by huge cables made up of thousands of strands of steel wire, ancient versions of the suspension bridge are said to have depended upon ropes of vines, bamboo or leather and eventually chains of iron.

Thomas Telford employed wrought-iron chains to support the 580 foot span of the Menai Straits Bridge. It was built in 1826 and was one of the first major suspension spans. Though renovated, it retains its original appearance and still carries traffic across the straits to the Island of Anglesey, Wales.

In 1832 Count Istvan Széchenyi founded a Bridge Association with the aim of building the first permanent bridge across the Danube at Budapest. The English engineer William Tierney Clark was chosen to design and build the Chain Bridge. The 600 foot span, officially known as the Széchenyi Bridge, was completed in 1849. It was rebuilt after the retreating German forces demolished it at the end of World War II and remains in use as a fitting centerpiece in that magnificent city. This photograph and my love affair with beautiful bridges both date from a visit to Budapest in 1985.

John Roebling perfected the process of spinning wire cables for use in long-span suspension bridges. His long and successful career in bridge building culminated in the design of the Brooklyn Bridge. Roebling was injured in a freak accident while surveying for the bridge footings and he died before construction was commenced.

His son Washington Roebling took over as chief engineer. It was under the supervision of the younger Roebling that the bridge was completed in 1883. At that time it was the longest span bridge in the world.

The Brooklyn Bridge's length has since been surpassed, but never its fame.

The Golden Gate Bridge stretches 4,200 feet between its soaring towers as it connects San Francisco with the Marin Peninsula to the north. It, too, was the world's longest span when it opened in 1937. That record was broken in 1965 at Verrazano Narrows and many times since, but nowhere in the world does a bridge combine more magnificently with its site.

San Francisco's other famous bridge, the San Francisco-Oakland Bay Bridge, was completed in 1936, a few months earlier than the Golden Gate. The suspension bridge has two end-to-end spans of 2,310 feet each. The sight of this bridge from Treasure Island at sunrise is very nearly as compelling as the view of its cross town rival.

CABLE-STAYED BRIDGES

The main cables of a suspension bridge are draped from the supporting towers and the deck is attached to the cables by means of vertical suspenders. In the cable-stayed form, the bridge deck is supported directly from the tower or towers by inclined stays. This technique was applied as far back as 1873 on the Albert Bridge over the Thames in London. Though it is also a suspension bridge, the principal support of its 450-foot span is borne by diagonal stays of wrought iron.

Faced with the task of rebuilding hundreds of bridges after World War II, the Germans employed the cable-stayed bridge design in response to the need for major spans which were economical in their use of scarce material. Since the 1950s the imaginative use of the cable-stayed form has added considerably to the aesthetic appeal of large and small bridges around the world. These are European examples.

Fleherbrücke, Germany 1979

Puente Sentinal, Spain 1990

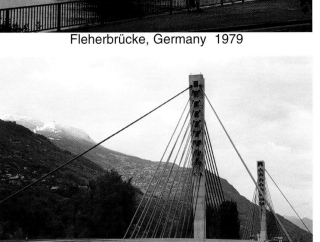
Pont de Chandoline, Switzerland 1990

Rheinbrücke, Germany 1978

In England the little Jackfield Bridge over the Severn was built in 1994. The people I talked to in nearby Ironbridge were not favorably impressed. They seemed to think that the imaginative use of the cable-stay form had gone too far.

The Annacis Bridge in British Columbia was completed in 1986. At that time its span of 1,526 feet was the longest in the world. At this writing it has already been surpassed at least ten times as the cable-stayed principle is applied to longer and larger structures.

A number of modern cable-stayed bridges in a variety of styles now cross the Mississippi River. These two examples carry highway traffic across the river between Illinois and Missouri.

Above is the Quincy Bayview Bridge. It was built in 1987 and has a span of 900 feet.

On the right is the Clark Bridge. It was completed in 1994 and the main span is 756 feet.

BRIDGES OF THE WORLD

he foregoing pages have demonstrated the various types of bridges. The next section will lead you to some of the wonderful places to view the best of the world's bridges.

Throughout the history of mankind, significant cities have been located almost exclusively on the principal waterways and accordingly are the sites of the great bridges of the world. Many other important bridges are located far from population centers but on traditional routes of trade or conquest. In either case, the process of seeking out these bridges leads one to some of the most fascinating places on earth.

Presented first are the bridges of North America. For obvious reasons there are no relics from Roman or medieval times here, but in the last two centuries the major rivers on the continent have been bridged by a wide variety of interesting structures.

Next are the bridges of Europe. The areas selected are those where the ravages of modern warfare have not decimated the fascinating ancient structures. Great Britain, France, Italy and Portugal provide wonderful examples of what can be in store for the traveler who is alert to discover the beauty of bridges.

BRIDGES OF NORTH AMERICA

North America is so richly endowed with notable bridges that it is difficult to narrow the selection to a manageable number. In order to give a sense of historical if not geographical continuity, those chosen for presentation here are arranged in chronological order.

Massive bridges of stone are not frequently encountered in America, but the Starrucca viaduct at Lanesboro, Pennsylvania is an outstanding exception. Built in 1848, it was the largest of its era and it still carries modern rail traffic 100 feet above Starrucca Creek.

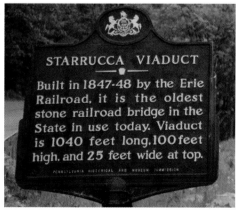

When it was completed in 1849, Charles Ellet's 1,010 foot suspension span in Wheeling, West Virginia was the longest in the world. In 1854 it was destroyed in a windstorm. It was immediately rebuilt and has had extensive renovations over the years. The stone towers are the only remnant of Ellet's original construction.

The Cincinnati Suspension Bridge, completed in 1867, was designed by John Roebling who is best remembered as the father of the Brooklyn Bridge. Reaching 1,057 feet across the Ohio River, when it was built it was another *longest span in the world*. Carefully maintained, it still carries its share of the traffic between Cincinnati, Ohio and Covington, Kentucky.

It was not easy to locate Wire Bridge Road in New Portland, Maine, but the reward was the opportunity to see and photograph this unique structure with its shingle clad wooden towers. There is some uncertainty as to the bridge's age, but it was probably built around 1868. It is remarkably well maintained and still in use.

The Eads Bridge crosses the Mississippi River at St. Louis with three unique tubular arch spans of 502, 520 and 502 feet. Intrigued for years by the story of James Eads and his monumental bridge, I was thrilled to find that it lived up to all my expectations. Completed in 1874, this pioneering structure was the only bridge that Eads built.

Pittsburgh, Pennsylvania is a city of bridges. The Smithfield Street Bridge, built in 1883, is the oldest survivor of the city's major historic spans. The unusual lenticular trusses support two spans of 360 feet each. The bridge was designed by Gustav Lindenthal (1850 - 1935) whose long career in bridge design was highlighted by New York's Hellgate Bridge. (See page 55.)

The huge Poughkeepsie Railroad Bridge, Poughkeepsie, New York carried rail traffic 212 feet above the Hudson River from 1888 until it was taken out of service in 1974. The future of the bridge is uncertain, but for now it remains a picturesque landmark.

Sometimes it can be extremely difficult to find a vantage point for a good view of a bridge. This is especially true of railroad bridges which are often located in remote areas away from access by road. The Kinzua Viaduct, built in 1900 in northern Pennsylvania, is an exception. As the centerpiece of Kinzua Bridge State Park, it invites close inspection. The dizzying view from the deck, 304 feet above the valley floor, is not for the timid.

New York City is a candy store for bridge lovers, especially those who admire suspension bridges. The Manhattan Bridge which spans the East River just upstream from the Brooklyn Bridge was completed in 1907. The pleasingly designed towers support a 1,470 foot span. It is difficult to find a New York bridge more visually appealing than the Brooklyn, but I consider the Manhattan a close second.

High Level Bridge, Lethbridge, Alberta, Canada. The stone marker at the east end of the bridge tells the story:

"RISING 307 FEET FROM THE VALLEY FLOOR THIS BRIDGE OVER WHICH RUNS THE CANADIAN PACIFIC RAILWAY THROUGH THE CROWS NEST PASS SPANS THE VALLEY IN ONE MILE AND FORTY SEVEN FEET. COMPLETED IN 1909 IT IS THE LONGEST HIGHEST BRIDGE OF ITS TYPE IN THE WORLD."

At Ferndale in Northern California, this concrete arch has carried traffic over the Eel River since 1911. When this picture was taken the river bed was mostly dry gravel, but when the spring floods rush out of the redwood forests, the broad channel is a raging torrent. Local residents have so far successfully resisted suggestions to replace their narrow old Fernbridge with a nice wide safe boring modern bridge.

Nicholson is a quiet little Pennsylvania village dominated by the Tunkhannock Creek Viaduct. Completed in 1915, totalling 2,375 feet in length, this immense concrete structure still carries the railroad 240 feet above stream level.

The Old Trails Bridge at Topock, Arizona was built in 1916. Its 592 foot arch carried the famous Route 66 across the Colorado River until it was replaced in 1948 and converted to use as a gas pipeline bridge.

On August 29, 1907, the huge partially completed cantilever bridge across the St. Lawrence at Quebec collapsed in a pile of twisted steel, carrying 75 workmen to their deaths. Pont Quebec, pictured here, was redesigned and finally completed ten years later, but not without additional mishap. The suspended central truss fell into the river during erection in 1916, killing an additional 11 workmen. At 1,800 feet, the bridge remains the longest cantilever span in the world.

The massive Hellgate Bridge carries rail traffic over the East River in New York. When it was completed in 1917, its 977 foot steel span was the world's longest arch bridge. Gustav Lindenthal was the designer. He was assisted by two young engineers, Othmar Ammann and David Steinman, who went on to achieve fame in their own illustrious bridge building careers.

The prime target for my bridging trip to Maine was this truly unique bridge which connects Bailey Island to the mainland. The roadway is supported by huge granite slabs assembled in an open cribbing which allows the strong tidal flow to pass through the 1,120 foot long causeway. Built in 1928, it is locally called the Cribstone Bridge.

THE WORLD'S HIGHEST SUSPENSION BRIDGE
Royal Gorge, Colorado

Most bridges are built for utilitarian purposes. This one was built in 1929 to attract tourists. It stretches 880 feet across the spectacular gorge, 1,053 feet above the Arkansas River.

It attracts tourists.

The Mid-Hudson Bridge at Poughkeepsie, New York was completed in 1930 under the direction of another American bridge building giant, Ralph Modjeski. It is widely admired for the artistic design of its towers. The 1,500 foot span is just downstream from the Poughkeepsie Railroad Bridge (page 47).

Speaking of artistic towers, David Steinman's St. Johns Bridge in Portland, Oregon demonstrates his flair for style. The 1,207 foot span was built in 1931. Interestingly, it has been cited by some as an example of tasteful design and by others as an example of inappropriate ornamentation. I think it is a classic!

New York's George Washington Bridge, designed by Othmar Ammann, was, in 1931, a giant step forward in both length and mass. Its span of 3,500 feet almost doubled the previous record held by Detroit's Ambassador Bridge. The striking appearance of the powerful towers is somewhat of an accident. They were to have been enclosed in masonry, but because of budget limitations the steel framework was left exposed.

1931 was a remarkable year for Othmar Ammann. The Swiss-born engineer achieved completion of two bridges which set world records. The George Washington Bridge (opposite) far exceeded all previous suspension bridges, while just a few miles away, the Bayonne Bridge established a new mark for arch structures. Also known as the Kill van Kull Bridge, the span of this 1,675 foot steel arch was not surpassed until the completion of the New River Gorge Bridge in 1977. (See page 67.)

The most famous bridge in the State of Washington achieved notoriety through its spectacular demise. The original Tacoma Narrows Bridge was completed in 1940. The slender span soon earned the name "Galloping Gertie" from its reputation for wild undulations in windy weather. Only four months after its opening, it dramatically tore itself apart during a moderate gale.

The second Tacoma Narrows Bridge, shown here, was completed in 1950. It spans the same 2,800 feet between towers but incorporates a wider, deeper and more rigid truss than did its ill-fated predecessor.

Michigan's Upper Peninsula is separated from the rest of the state by the Straits of Mackinac, five miles of open water subject to violent winds and daunting ice. Bridging the Straits was the crowning achievement in David Steinman's long and distinguished engineering career.

His Mackinac Bridge, completed in 1957, has a 3,800 foot main span suspended from towers that rise 552 feet above water level. Combining the main span with the two 1,800 foot side spans, the Mackinac* still ranks among the world's longest suspension bridges.

*Pronounced Mackinaw.

Kinnaird Bridge, built in 1965, crosses the Columbia River at Castlegar, British Columbia. I was impressed with its style, but was unaware of its pedigree until later when I learned that the highly regarded Italian engineer Riccardo Morandi collaborated in its design. Morandi's most famous structure is the giant cable-stayed Lake Maracaibo Bridge in Venezuela. It is on my list for a future bridging trip.

Most modern suspension bridges are built where the situation calls for a particularly long span. An interesting exception is this beautiful little 430 foot Orleans Bridge which was built in 1967. It is hidden away in the remote and scenic Klamath River Valley in Northern California.

When a new freeway was planned to cross the Willamette River harbor in Portland, Oregon, there was local pressure for a structure with an aesthetically pleasing design. The form chosen for the Fremont Bridge was a steel through arch. It carries eight traffic lanes on two decks, spanning 1,255 feet. Built in 1973, the top of its arch towers 381 feet above the river.

The New River Gorge Bridge in West Virginia is the longest arch bridge in the world. The 1,700 foot steel deck arch was completed in 1977. Four lanes of highway traffic are carried 876 feet above the surface of the New River.

For the bridge fan there is an appropriately elaborate interpretive center and viewing area at the site.

In an era when so many highway structures are either unattractive or at best inconspicuous, it is satisfying to find instances where modern bridge designs actually complement the beauty of a scenic location. This graceful steel through arch is certainly one of those instances. The bridge is at Roosevelt Dam near Phoenix, Arizona. It was built in 1990.

Bridge engineer Conde B. McCullough is the subject of a later chapter. One of his most admired bridges crossed Alsea Bay at Waldport, Oregon from 1936 until 1991. It was replaced after marine salts penetrated into the old bridge's concrete, rusting and weakening its reinforcing steel beyond repair. A concerted effort was made to design a new bridge which would approach the aesthetic appeal of the original. The result was this fine new concrete structure which incorporates a steel tied arch of 450 foot span. Compare it to its predecessor on page 137.

The Great River Bridge which crosses the Mississippi River at Burlington, Iowa was completed in 1994. The modern 660 foot cable-stayed span replaced a much less dramatic cantilever bridge which was built in 1917. The pier in the foreground is a remnant of the old bridge. When visiting Burlington, I was fortunate to catch a break in the murky sky and enjoy the illumination of the bridge in the warm light of the late evening sun.

The selection of color for a bridge can be quite controversial, and in many instances aesthetically inappropriate choices are made.

The striking red of the New Salt River Canyon Bridge contrasts sharply with its bypassed predecessor, but is quite at home in this colorful Arizona canyon. The old bridge was built in 1932 and the new one was completed in 1996.

BRIDGES OF GREAT BRITAIN

Few cities rival London for the opportunity it provides to view a concentration of interesting bridges. The Thames has been bridged here for at least 800 years, but the oldest remaining spans are from the 19th century. Waterloo Bridge, built in 1942, is sleek, functional and colorless. Pictured opposite are some of its more resplendent neighbors.

Lambeth Bridge

Southwark Bridge

Hammersmith Bridge

Vauxhall Bridge

Battersea (foreground) and Albert Bridges

Blackfriars Bridge

With its rows of shops and houses, Old London Bridge was probably the most famous of "inhabited" bridges. It was replaced in 1831, but Great Britain still has several interesting examples of bridges which support fortifications, shops, houses and chapels.

The newest of the examples pictured here, Pulteney Bridge at Bath, was completed in 1773. Designed by Robert Adam, it still carries two rows of shops across the River Avon.

The oldest is the Monnow Bridge at Monmouth. This fortified medieval bridge dates from 1272.

Lincoln's 16th century High Bridge is one of the few that still supports houses, four stories of houses.

The smallest, the Bridge House at Ambleside, was built around 1610 as a "folly." Its tiny two-story building is a tourist attraction now.

The building on this 14th century bridge at Bradford-on-Avon has at various times served as a chapel, a lock-up and a powder magazine.

There are so many attractive old stone bridges surviving in cities throughout Great Britain that it is very difficult to select the most picturesque. Scotland's Old Stirling Bridge (above) is a case in point. It dates from around 1400 and is now used as a footbridge only.

Bishops Bridge over the Wensum in Norwich dates from 1249. As quarry stone was not abundant in this part of England small stones called flints were used in the spandrel walls.

Attractive bridges of iron and steel are less common. The Old Road Bridge over the Wye at Chepstow, Wales was built in 1816. Its lacy cast-iron arches have a maximum span of 112 feet.

When the Forth Road Bridge was completed in 1964, some scoffers considered it to be just a viewing platform for the nearby Forth Rail Bridge. I think it is a thing of beauty in any light.

In Scotland attractive little stone arch bridges seem to be every-where. At Keith in Grampian, the Auld Brig is tucked away in a quiet park, bypassed and no longer in use. It was built in 1609.

At the top of a mountain pass in Argyl, Scotland is a pleasant stopping place which is aptly named Rest-and-be-thankful. The little stone arch there is Butterbridge, built in 1745.

The Bridge of Carr at Carrbridge, Scotland provides a remarkable demonstration of the strength, stability and beauty of the arch form. Built in 1717, all that remains of the 34 foot arch is this narrow ring of voussoirs. (Voussoirs are the wedge shaped stones that form the arch ring.) I felt perfectly safe walking over it.

After the rebellion in 1715 in Scotland, English General George Wade was given the task of building an extensive network of military roads to help bring the Highlands under control. Whether or not he was personally involved in their design, numerous bridges built by his forces survive today and are still called Wade Bridges.

The most elaborate is this bridge of five stone arches across the Tay at Aberfeldy. Built in 1733, the center arch has a 62 foot span.

Garvamore Bridge was built by Wade's men in 1731. The search for it took us up a minor road through some of the most beautiful country in Scotland.

Nearby at Glen Shirra we found another little Wade Bridge that has been bypassed by both the river and the road.

The 18th and 19th centuries in Britain saw the development of improved transportation in three phases: the canals, the railroads and the highways. All three provided unprecedented opportunities for bridge engineers. Of the many famous engineers who flourished in the environment, none was more prolific than Thomas Telford (1757 - 1834). Many of his varied designs are still in use. Two of my favorites appear in earlier chapters, Menai Straits (page 31) and Betwys-y-Coed (page 16).

The first of many Telford bridges in Scotland, the Tongland Bridge, was built in 1808. Its castellated 112 foot stone arch spans the River Dee.

At Craigellachie, also in Scotland, the castellated stone towers are combined with a graceful cast-iron arch. The 150 foot span was completed in 1814.

In Wales the Chirk Aqueduct, built in 1801, carries the Ellesmere Canal 70 feet above river level in the Ceiriog valley.

Telford's most flamboyant use of castellation is his attempt to blend the architecture of his 1826 Conwy Bridge with that of the adjacent Conwy Castle.

Closing the chapter on Great Britain are two icons of Victorian age engineering, both located in Scotland. On the left is the Connel Ferry Bridge. Though it was built in 1903, I think it has a strikingly modern look.

But to me, the most impressive bridge in all of Great Britain is the Forth Rail Bridge (opposite).

In 1879 the recently completed Tay Bridge collapsed in a storm and carried a train into the Firth of Tay with the loss of all of the 75 people aboard. The designer of the Tay Bridge, Sir Thomas Bouch, had already begun construction of an even more ambitious bridge at the Firth of Forth. In the wake of the disaster, Bouch was discredited and his plans were scrapped. A new design for the Forth Bridge was prepared and executed by Sir John Fowler and Benjamin Baker. In reaction to the recent tragedy, their massive cantilever design was calculated to restore public confidence. Upon completion in 1890, its two spans of 1,710 feet broke all the world records.

BRIDGES OF FRANCE

Paris, like New York and London, is a bridge lover's paradise. The oldest remaining bridge across the Seine is the Pont Neuf. When it was completed in 1606, it was unique among the bridges of Paris in that it was designed exclusively to carry traffic rather than to support shops and homes. Actually, until their removal in 1850, it did support rows of makeshift shops and market stalls.

After the Pont Neuf came a series of elegant stone arch spans. The Pont Royal was completed in 1687. With its five elliptical arches and relatively narrow piers, it set a new standard that Paris saw repeated in the ensuing centuries.

As recently as 1857, with the construction of the Pont Saint Michel, the style, though more ornate, was essentially the same.

Passerelle de Billy is a graceful steel arch footbridge, built in 1899 for the Exposition of 1900. For my purpose it provides the opportunity to include a picture of the Eiffel Tower.

Paris's most ornate bridge is the Pont Alexandre III. Completed in 1899, it too was built on the occasion of the Paris Exposition and was named in honor of the Russian Czar. Its cast-iron arch spans the Seine in a single leap of 353 feet.

The Pont d'Avignon is the subject of legends and songs. Legend tells us that it was built by a lowly shepherd boy who was eventually to be canonized as St. Bénezet. French children sing the song "Sur le pont d'Avignon." When completed in 1188 there were 20 or 21 stone arches stretching across the Rhone. Now only four arches remain, but the bridge has retained both its fame and its beauty.

The Pont sur la Bévéra was also built in the 12th century. It is the centerpiece of the lovely town of Sospel, near the Italian border northeast of Nice. My brief visit there was made memorable by the friendly villagers who went out of their way to help me find a good vantage point.

The Pont de Langlois is a wooden bascule drawspan dating from around 1820. It achieved fame as the subject of a painting executed in 1888 by Van Gogh. In 1962 the bridge was removed from its original site to the present location near Arles where it is preserved as a tourist attraction.

When the angle of the morning sunlight is just right and the strollers are silhouetted perfectly against the lake and the mountains, a simple footbridge provides a picture every bit as romantic as its name. The bridge at the beautiful alpine town of Annecy is the Pont des Amours.

BRIDGES OF SPAIN

To discover the widest array of bridges, visit Spain. There you will find an abundance of interesting spans from every era, Roman, Moor, medieval, modern and *very* modern.

All that remains of some of the Roman bridges are ruins of half buried arches, like this example at Aznalcazar. At Carmona (below) the bridge once used by Roman legions is now a path for grazing sheep.

The Puente Romano at Mérida (opposite) is much more sound and still in use. Built about 25 B.C. its multiple arches reach over 2,000 feet across the Guadiana River, ranking it as the longest of all remaining Roman bridges.

The Puente Romano in Salamanca was built around 100 A.D. during the reign of Emperor Trajan. The massive granite spans on the left are the original Roman construction.

In the 17th century, eleven arches were added (on the right in the picture above) extending the bridge to its present length of 1,168 feet.

It is difficult, even for the experts, to determine the age of many of Spain's ancient bridges. Puente Viejo in Ávila looks Roman. It is still in use as a footbridge while the modern traffic is carried by the adjacent Puente Nuevo.

Opposite experiences in bridging can be equally satisfying. One is the chance discovery of a spectacular bridge that was heretofore unknown to you. The other is to finally behold a bridge which you have researched and long anticipated. In the latter category, pictured on these pages, are Toledo's historic bridges over the Tagus River. Both lived up to expectations.

The older of the two medieval bridges is the Puente de Alcántara. It rests upon Roman foundations. It was rebuilt by the Moors in the 9th century and again by the Christians in the 13th century.

According to legend, the builder of the Puente San Martin told his wife of his fear that an error in his design would cause the central arch to collapse when the temporary wooden centering was removed. That night the woman set fire to the centering. The fire destroyed the bridge and thus her husband's reputation was preserved. After her husband successfully rebuilt the bridge, she confessed her act to the Archbishop. It is said that instead of punishing her he commended her loyalty. Whether or not the legend is true, the bridge, built around 1200, still stands.

The thrill of a surprise discovery is exemplified by the bridge over the Rio Pusa. Returning to Toledo after a day of bridging in the countryside to the west, we happened upon this little canyon near the village of Santa Ana de Pusa. So far, I have been unable to determine the age or even the identity of the elegant brick and stone arch bridge.

Another chance discovery was made at Castillobel on our way back to Barcelona after a visit to Montserrat. While I clambered down the river bank to get my pictures, my wife and daughter visited with the friendly local shopkeeper. His story was that the medieval bridge was built by an evil baron who extracted heavy tolls for its use.

My first visit to Ronda was on an Easter Sunday morning. The drama of the religious procession combined with the spectacle of the Puente Nuevo made the experience unforgettable. The bridge, built in 1788, connects the old and new sections of the city which is bisected by the gorge of the Guadalevin.

The vibrant colors of the ceramic balustrades accentuate the graceful lines of the ornamental bridges at the Plaza de España in Seville. Built in 1929 as part of the Ibero-American Exposition, they are often featured in tourist publications depicting the charms of Spanish architecture.

As stated in the introduction to this chapter, Spain is blessed with not only an abundance of ancient bridges, but also with the very modern. Here are the *very* modern. Discovering the Puente Lusitania in Mérida was one of my most exciting moments in bridging. This 1991 creation of Spanish-born architect/engineer Santiago Calatrava is positioned just downstream from the Roman bridge pictured on page 95.

The most spectacular example of Calatrava's work is the Puente Alamillo in Seville. The 656 foot cable stayed span was built in 1992. The inclined pylon soars 465 feet above the deck. There are no back stays; the weight of the concrete filled steel tower supports the deck. Hopefully, Calatrava's imaginative and controversial designs have stimulated others to create bridges which are works of art.

BRIDGES OF ITALY

The Romans weren't the first bridge builders. Even in Italy, the Etruscans preceded them. But the Romans' widespread achievements in bridge construction were not surpassed for over 1,000 years. Many Roman spans were built of timber, but the surviving structures are the magnificent stone arches. In Rome the Tiber River is spanned by a number of remarkable bridges, many of Roman origin. The best preserved of these is the Ponte Fabricio.

The Ponte Fabricio (opposite) was built in 62 B.C. and still bears the inscription crediting L. Fabricius with its construction. The Ponte Sisto (below) also crosses the Tiber in Rome. Its dedication is dated MCCCCLXXV. The similarity of these two bridges is a striking demonstration of the lack of development of the art of bridge building in the intervening 1,500 years.

The Pont St. Martin in the town of the same name in the Aosta valley in northern Italy boasts the longest span of any of the Roman bridges. The 120 foot arch was built as part of a military road. The experts disagree about the exact time of its construction, but estimates place the date from 140 B.C. to 25 B.C. In any case, old.

At Rimini, construction commenced on this ornate bridge during the reign of Emperor Augustus and was completed in 20 A.D. during the reign of Emperor Tiberius. Though often referred to as Ponte d'Augusto, the local signs direct the busy traffic to Ponte de Tiberio. Considered by many to be the most beautiful of the Roman bridges, it was admired and copied in the 16th century by Palladio.

The Ponte dell'Abadia is located at the site of the ancient Roman town of Vulci. It is believed to date from about 90 B.C. The bridge appears to be built upon older foundations, possibly Etruscan. The impressive arch carries a narrow footpath 114 feet above the River Fiora.

The protruding stones on the face of the arch are a common feature of masonry bridges. They probably supported the temporary falsework which was employed to bear the weight of the arch during construction.

The picture on the right was taken from the very interesting Etruscan museum which is adjacent to the bridge.

In the first century B.C. the Via Salaria was a major road from Rome north through Ascoli Piceno to the Adriatic. The Ponte Quintodecimo is one of many fine Roman bridges surviving along the route. Its well-preserved travertine arch has a span of 56 feet.

The indentations in the arch face probably performed the same function as the protruding stones on the opposite page. They would serve as anchor points for the timber falsework which propped up the arch until the voussoirs were all in place and the arch became self-supporting.

Near the Umbrian town of Acquasparta, the abandoned medieval church of San Giovanni de Butris stands astride the remaining arches of a Roman bridge. In 25 B.C. a branch of the Roman Via Flaminia crossed a stream here. The stream has long since changed its course and the arches provide a solid foundation for the little church.

For a thousand years after the decline of the Roman Empire, no important bridges were built in Italy. The graceful Ponte Maddalena, near Bagni di Lucca, was constructed in 1317. This is one of the many bridges which share the name Ponte del Diavalo--Devil's Bridge.

All of the bridges across the Arno in Florence were destroyed by the retreating German army in 1944. All except the most famous, the Ponte Vecchio. It was built by Taddeo Gaddi in 1345. The bridge is lined on both sides with shops which originally housed butchers, grocers and blacksmiths, but since the 16th century have been occupied by gold and silversmiths. The upper gallery is a private passage that connects palaces on opposite sides of the Arno.

Florence's Ponte Santa Trinita, built in 1567, destroyed in 1944, was lovingly restored by 1958. Most of the original stonework was recovered from the river and the lost pieces were faithfully replicated from the original quarries.

The last piece recovered was the head which was missing from this statue. After a successful international publicity campaign asking, "Have you seen this woman?" the missing portion was located in Greece and restored to the statue with great fanfare.

The tourist guides promote Verona as the city of Romeo and Juliet. For those of us who admire the beauty of bridges, it is the site of the Ponte Scaligero and the Ponte de Pietra. The Ponte de Pietra was constructed by the Romans, probably in the 1st century B.C. It has been rebuilt many times, most recently in 1959 after its destruction at the end of World War II, but always retaining its original Roman form.

The Ponte Scaligero (also known as the Ponte Castelvecchio) dates from 1356. It crosses the river Adige directly into the Scaliger Castle and its crenelated defensive walls are a continuation of the castle's fortifications. The span closest to the castle is 160 feet, one of the longest medieval masonry arches.

On the right is the defender's view from the bridge.

Venice

Although Venice is a city of hundreds of charming bridges, only two are well known. The Rialto Bridge (above) was built in 1592 by Antonio da Ponte. Tradition has it that his design was selected in preference to plans proposed by, among others, Michelangelo and Palladio (both of whom died before 1592).

The Bridge of Sighs (right) connects the Doge's Palace with the prison. It takes its name from the sighs of the prisoners about to be incarcerated or executed. Its design was also attributed to da Ponte, although his nephew Antonio Contino is credited with its construction.

At Bassano, the alpine fury of the River Brenta has been destroying bridges since the 12th century. In 1568 the town fathers selected the famous architect Andrea Palladio to design a replacement for the bridge which was swept away the prior year. The il Ponte degli Alpini has been faithfully rebuilt to Palladio's distinctive plan after the ravages of high water in 1748, 1813, 1945 and 1966.

Although there are many fine modern bridges in Italy, I haven't found any that compare in beauty to the ancient spans. Consequently, the newest bridge featured in this chapter is the Treponti, built in 1634. It is one of a group of whimsical little spans in Comacchio that the guide books describe as "operetta" bridges.

BRIDGES OF PORTUGAL

Like Italy and Spain, Portugal has a wealth of well-preserved ancient bridges dating back as far as Roman times. There are also some relatively modern spans worthy of note. The most unusual is the Ponte Luiz I in Porto. It was built in 1885 by Thoephile Seyrig who was associated with Eiffel in the construction of the nearby Ponte Maria Pia (see page 17). This arch with a span of 566 feet carries traffic on two levels over the Douro River.

Porto is the site of four monumental bridges, the 19th century iron arches of Seyrig and Eiffel contrasting with Edgar Cardoso's 20th century concrete creations. The 890 foot span Ponte Arrábida (below) was the world's longest concrete arch when built in 1963. The latest, completed in 1991, is the starkly attractive Ponte Sâo Joâo which has a main span of 820 feet. Eiffel's Ponte Maria Pia is in the background.

When the Ponte 25 de Abril was constructed in 1966, its 3,323 foot main span was the longest in Europe.

If my scheduled time in Lisbon had been a little more generous, I would have gotten a better picture. Good sense required that instead of waiting for the fog to burn away, I should keep my lunch date with my wife at the restaurant atop the Elevador de San Justa. Bridges aren't the only pleasures of travel.

Not all of Gustave Eiffel's works are as spectacular as the Eiffel Tower or as graceful as the Ponte Maria Pia. The Ponte Internationale at Valenca do Minho on the Spanish border was built by Eiffel in 1885. Rail traffic is carried on top, above the roadway which runs through the wrought-iron lattice girder.

Clapper bridge, causeway, flag bridge or slab, whatever it is called, is one of the most basic bridge types. It is simply one or more stone slabs placed on supporting rock piers. Portugal provides a variety of examples .

The footbridge (above) over the River Vez at Arcos de Valdevez is paved with six-foot-long slabs of stone.

Some bridges are quite crude, but some have carefully cut and fitted stones like this ageless span which crosses a stream in a little mountain meadow near Tarouca Alcobaca.

On the advice of the hotel clerk in Chaves, we altered our plans to include a stop at Carvelhelhos. We were rewarded with this pair of bridges. The arch bridge is the Ponte Pedrinha from the 13th century. The adjoining causeway is called Ponte Romana, but is probably not old enough to be Roman. The stone slabs that pave the causeway are up to six feet long.

Picturesque medieval bridges are still in use throughout Portugal. Mirandela's well-maintained Ponte de Mirandela had just been washed by a heavy spring rain shower when the sun emerged from the clouds to enhance this photograph. Opposite are the Ponte da Barca at the town of the same name, and (below) the Ponte do Porto at nearby Figueiredo.

The village of Idahna a Velha is the site of another Ponte Romana, but it is not necessarily Roman. There is no doubt that the Romans had established a network of roads throughout Portugal and Spain during the reign of Augustus (27 B.C. - 14 A.D). Historians and archeologists have identified the major Roman bridges, but for minor bridges such as this, the picture is less clear. The dating is often confused by reconstruction and repairs. The pointed second arch, for instance, is not in keeping with Roman building technique or, for that matter, with the other arches of this bridge.

Ponte da Misarela is another of the many medieval bridges with a devil's legend. A fugitive fleeing from his pursuers found himself trapped at the edge of this mountain torrent with no way to cross. The devil offered to provide a bridge in trade for his soul. The deal was made and the devil destroyed the bridge after the man's escape. Repentant, the man sought the assistance of his priest to regain his soul. He dealt with the devil again for a bridge over the chasm, but this time the wily priest concealed himself nearby and blessed the span before the devil could destroy it, thereby preserving it for all time.

This chapter started with the statement that Portugal has a wealth of well-preserved ancient bridges dating back as far as Roman times. As interesting as the modern and medieval spans are, none surpasses the beauty of the Roman bridges.

The bridge at Vila Formosa carried the Roman road over the River Sêda. This was the same Roman road that passed through the Spanish city of Mérida. Compare the striking similarities of the configuration shown on these two pages with that of Mérida's Puente Romano pictured on page 95.

Modern highway traffic is still carried by this bridge which has been described as the best preserved Roman bridge in Portugal. The Vila Formosa Bridge is especially appealing because its isolated location renders it easily approachable and eminently photographable.

Locally the Ponte Quinhentista at Portagem is considered to be a Roman bridge, though it is not described in the authoritative literature on Roman bridges. Its location on the route from Merida to the west and its style of construction certainly lend credence to the local claim. I recall spending a few minutes huddled for shelter under one of the approach arches waiting out a temporary downpour just before I took this picture.

Just down the road from the Portuguese hilltop town of Segura, the Elja River forms the border with Spain. Here the river is crossed by the Ponte de Segura, a Roman bridge of generous proportions. It is often compared with the much larger nearby Alcántara bridge (page 3) and is probably from the same period, 104 A.D. The mixture of construction materials indicates that the original Roman piers and arches have been augmented by less elegant stonework in more recent times.

PERSONAL FAVORITES

s the focus of my interest in bridges is based primarily upon their aesthetic appeal, it is only natural that I am often asked to identify my favorite bridge. There is such a wide variety of bridge types and so many wonderful examples of each that I find it impossible to come up with a single answer. I do, however, have some favorite categories.

The bridges in the previous chapters were organized by construction material and then by geographical location. Within each of these classifications I have tried to present the most interesting of the photographs in my collection. But I have held aside until now three groups of pictures which certainly do qualify as favorites. First, the pride of my own state, the unique and varied bridges of Conde B. McCullough.

Oregon's Conde B. McCullough Bridges

Conde B. McCullough (1887 -1946) was an engineer in the right place at the right time. Born in South Dakota and educated in Iowa, he left his post as a Professor of Engineering at Oregon State College in 1919 to become the state bridge engineer. It was a time when the need for expansion of the highway infrastructure furnished the occasion for the design of many new bridges. With the depression in the 1930s, federal funds became available for ambitious public works. These factors, combined with McCullough's creativity, provided the opportunity and the means for the building of an unparalleled variety of unique and attractive spans. The most famous of his bridges are links in the Oregon coast highway which was completed in 1936.

The Alsea Bay Bridge at Waldport (opposite) was my choice for the cover picture of my earlier book. Its total length of 3,011 feet ranked it as McCullough's largest concrete structure, and it was widely regarded as one of the finest concrete bridges in America. Unfortunately, it suffered serious structural deterioration as a result of the corrosion of its steel reinforcement. It was replaced in 1991 by a new bridge. (See page 69).

While most of McCullough's bridges are executed in reinforced concrete, two of the most notable spans are primarily steel. Both of these examples include concrete approach arches typical of his style, but the main span is steel.

The Yaquina Bay Bridge at Newport is one of the coast highway bridges which were completed in 1936. The bridge reaches 3,223 feet across the bay. Its soaring channel span is a 600 foot steel through arch flanked by two steel deck arches of 350 feet each.

The McCullough Memorial Bridge crosses Coos Bay on the southern Oregon coast. It is the largest of the bridges completed in 1936. Its overall length is 5,305 feet and its steel cantilever truss provides a clear span of 793 feet. The bridge was posthumously dedicated to McCullough in 1947.

Both Yaquina Bay and Coos Bay are visited by ocean going ships and therefore require substantially more channel width and vertical clearance than the other coastal bridges.

The variety of designs McCullough executed in reinforced concrete is demonstrated in this selection of Oregon coast highway bridges. The two tiered arches of the 1936 Cape Creek Bridge (above) give it the grace of a Roman aqueduct. Opposite, the 1927 Ben Jones Bridge and the 1931 Cummins Creek Bridge are both concrete deck arches. Tenmile Creek Bridge was one of three reinforced concrete tied arch bridges built along the coast road in 1931, the first of this type in America. The Siuslaw River Bridge at Florence combines concrete tied arches with a steel double bascule lift span. It was another of the large coastal spans completed in 1936.

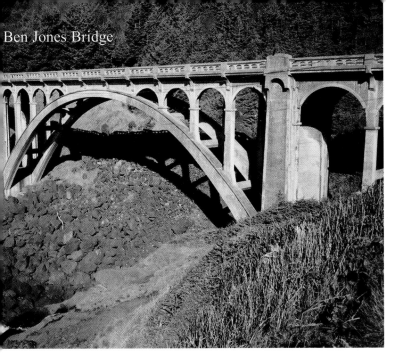

Ben Jones Bridge

Cummins Creek

Tenmile Creek

Siuslaw River

The Oregon coast wasn't the only area of the state that benefited from McCullough's work. The gorge of the Crooked River in Central Oregon is the site of this 330 foot span steel arch, built in 1926. The highway is carried 295 feet above the stream below.

Opposite are more examples from all corners of the state.

Fifteenmile Creek Bridge, in rural Wasco County in Eastern Oregon was built in 1925.

The Rock Point Bridge spans the Rogue River in Southern Oregon. It was built in 1920.

The Hood River is crossed by Tucker's Bridge near the city of Hood River in the Columbia Gorge. McCullough built this span in 1932.

Oswego Creek Bridge is the nearest to populous Portland of any of McCullough's bridges. But it is seldom noticed through the heavy foliage.

McCullough designed two interesting versions of the half through arch. The Caveman Bridge at Grants Pass (above), built in 1931, is made up of three 150 foot reinforced concrete spans. At Oregon City (below) the 360 foot span appears to be a concrete arch but is actually steel encased in a protective layer of sprayed-on gunite. It was built in 1922 replacing a suspension bridge of 1888. It has in turn been supplanted in 1970 by the freeway bridge in the background.

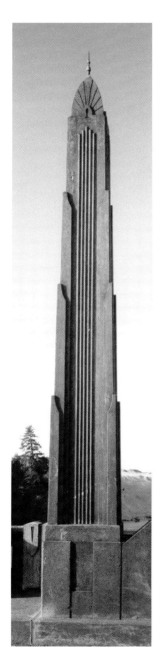

In contrast to the unadorned concrete and steel bridges that are the modern standard, Conde McCullough's works were noted for the interesting ornamentation he employed. "Noted" includes both criticism and praise.

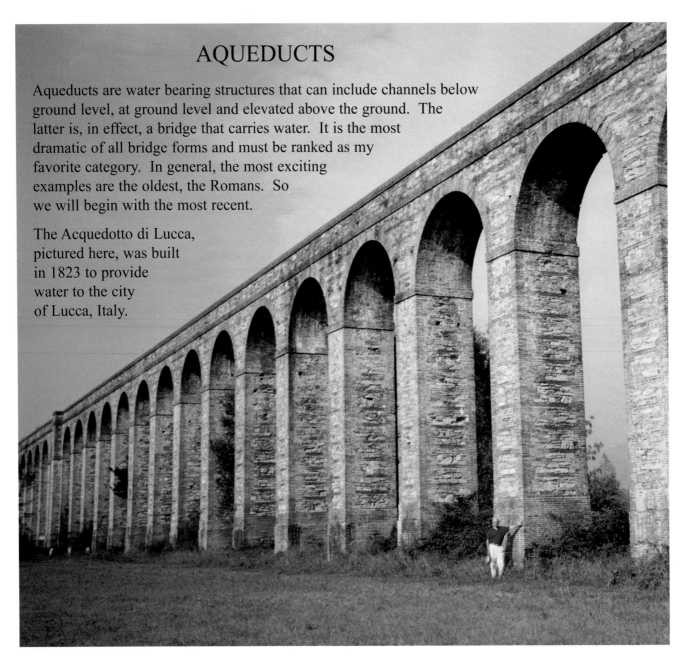

AQUEDUCTS

Aqueducts are water bearing structures that can include channels below ground level, at ground level and elevated above the ground. The latter is, in effect, a bridge that carries water. It is the most dramatic of all bridge forms and must be ranked as my favorite category. In general, the most exciting examples are the oldest, the Romans. So we will begin with the most recent.

The Acquedotto di Lucca, pictured here, was built in 1823 to provide water to the city of Lucca, Italy.

Returning from an exhilarating day of sightseeing at Granada, I was astounded to come across this four tiered aqueduct near Almuñécar on Spain's Costa del Sol. There is also a Roman aqueduct at Almuñécar, but this structure is said to be much more recent, probably 18th century.

Lisbon's massive Aqueduto das Águas Livres was built in 1748 by Manuel de Maia. It enters the city with this huge gallery of 35 arches over the Alcântara Valley. The central arch has a maximum span of 95 feet and reaches a height of 213 feet above the valley floor. It is claimed to be the world's largest stone arch. At one time pedestrians could enjoy the vertiginous view from a footpath alongside the enclosed water channel, but alas, no more.

In Portugal, a number of exciting aqueducts survive from the 16th, 17th and 18th centuries.

On the left, the Aqueduto do Vila do Conde dominates the skyline of the city of Vila do Conde. Completed in 1714, its 999 arches extend for a length of three miles.

At Elvas, below, the Amoreira Aqueduct was started in 1530 by Francisco de Arruda, but not completed until 1622. At its highest point, the four tiers are over 100 feet tall.

Above the Portuguese city of Tomar is the magnificent Convento de Cristo. Since 1614, water has been carried to the convent by the equally impressive Aqueduto de Pegões*. Here the huge structure sweeps across the valley in a colonnade of 300 arches. The picture on the opposite page shows the unusual combination of pointed gothic arches and semicircular arches in the double tiered portion. The view from the top reveals the partially uncovered water channel flanked by a footpath.

* Our friendly young waiter at dinner made a valiant effort to teach me the pronunciation of *Pegões*.

The Usseira Aqueduct (above) was constructed in the 16th century to supply water to the walled city of Óbidos, Portugal.

The imposing aqueducts on the opposite page are also located in Portugal. On the left, at Serpa, water was lifted from an underground channel and carried into the city on this row of arches that form part of the city wall. The 17th century lifting device employs an endless chain of clay pots.

On the right is the Aqueduto da Água Prata at Évora. It brought water from springs located 11 miles from the city. It was designed by Francisco de Arruda and completed in 1537.

The magnificent Ponte della Torre at Spoleto, Italy is not only an aqueduct, but also a bridge. Its ample footpath allows a breathtaking view of the Umbrian countryside from 266 feet above the valley floor. The original structure here has been attributed to the Duke of Spoleto in 604, but what we see today is a reconstruction from the 14th century.

The Romans were not the originators of the aqueduct, but they mastered the technique of supplying fresh water wherever their legions spread Roman civilization. The ancient city of Rome itself was served by eleven major aqueducts bringing water from as far away as 57 miles. Many of the far-flung Roman aqueducts are now fragmentary, but there are also magnificent examples which are intact after as long as 2,000 years.

Pictured here is the Acueducto de Segovia in the city of Segovia, Spain. It was built in the 1st century A.D. and still carries water. The elegant double tier of arches towers 93 feet above street level, dominating the center of the city.

Near Avignon, France is the most renowned aqueduct of all. Widely acclaimed as one of the world's most beautiful bridges, the Pont du Gard was built in 18 B.C. It was a part of a 31 mile aqueduct which supplied water to the Roman city of Nemausus, now known as Nîmes. Its dimensions are prodigious, even for the Romans. Its central arch has a span of 80 feet and its three tiers soar 160 feet above the river Gard, making it the tallest known Roman bridge. As you can see, it is also in a charming and very popular site.

THE LITTLE WOMAN

This book is dedicated, as was my first book,

To My Wife, Kathy

For her support and encouragement in spite
of having to wait patiently at every bridge.

In olden times (my generation) the term "the little woman" was intended as an endearing reference to one's wife. While in this day and age it may no longer be acceptable, it accurately defines one of the roles that my wife fulfills as she waits patiently at every bridge.

In order to reflect the scale of a bridge, it is often appropriate to pose my companion near, under or upon the span. The collection of pictures that follows includes some of my favorites of "the little woman."

The enormity of the Bixby Creek Bridge on California's beautiful Big Sur coast is brought into focus by Kathy's pose on the bench above the towering pier.

This 330 foot reinforced concrete arch was built in 1933. One of its designers was Charles Purcell whose main claim to fame was the San Francisco-Oakland Bay Bridge.

On the left, the awesome scale of the Ponte della Torre at Spoleto, Italy is apparent when you spot my tiny wife sitting on the stone bench in the uppermost arch.

Kathy may appear bored, "waiting patiently" beside this bridge in Ascoli Piceno, Italy, but she is actually enjoying the sunshine and savoring the memory of Ascoli's special stuffed olives.

At the Ponte Mallio in Cagli, Italy (above), Kathy's presence accentuates the generous dimensions of the individual blocks of stone that the Romans employed in the construction of this bridge. The bridge dates from the 1st century B.C. The model is younger.

The same comparison is apparent at San Giovanni de Butris (below). This bridge surmounted by a church is also pictured on page 112.

Kathy is dwarfed by the huge concrete approach arches of the Hellgate Bridge in New York. Refer to page 55 for a more conventional view of the muscular steel arch of this railroad bridge.

Many bridges have unsavory undersides which I wouldn't invite Kathy to share with me. I have slid down steep embankments, crashed through thorny underbrush, ignored garbage, filth and the friendly warnings of concerned observers to find my way to a vantage point under a bridge. But in Seattle, the whimsical Giant Bridge Troll that lurks beneath the Aurora Bridge is a delight.

INDEX